POSTMODERN ENCOUNTERS

Marshall McLuhan and Virtuality

Christopher Horrocks

Series editor: Richard Appignanesi

ICON BOOKS UK

TOTEM BOOKS USA

Published in the UK in 2000
by Icon Books Ltd., Grange Road,
Duxford, Cambridge CB2 4QF
email: info@iconbooks.co.uk
www.iconbooks.co.uk

Published in the USA in 2001
by Totem Books
Inquiries to: Icon Books Ltd.,
Grange Road, Duxford,
Cambridge CB2 4QF, UK

Distributed in the UK, Europe,
Canada, South Africa and Asia
by the Penguin Group:
Penguin Books Ltd.,
27 Wrights Lane,
London W8 5TZ

In the United States,
distributed to the trade by
National Book Network Inc.,
4720 Boston Way, Lanham,
Maryland 20706

Library of Congress catalog
card number applied for

Published in Australia in 2001
by Allen & Unwin Pty. Ltd.,
PO Box 8500, 9 Atchison Street,
St. Leonards, NSW 2065

Text copyright © 2000 Christopher Horrocks

The author has asserted his moral rights.

Series editor: Richard Appignanesi

ISBN 1 84046 184 5

Typesetting by Wayzgoose

Printed and bound in the UK by
Cox & Wyman Ltd., Reading

Introduction: Saint McLuhan

Man, he understood the Internet. He was the Internet in the sixties. The world's just finally caught up to him. He was an internet in the sense he was in touch with the entire globe. . . . He was wired long before the editors of Wired *magazine were born. This man was truly wired.*[1]

Robert Logan

. . . the Christian concept of the mystical body – all men as members of the body of Christ – this becomes technologically a fact under electronic conditions.[2]

Marshall McLuhan

The arrival of the computer-driven information revolution, along with its strange geographies of cyberspace, virtual reality, the Internet and the Web, has for some people reinvigorated the writings and ideas of Marshall McLuhan. For many, his name is synonymous with his phrase 'the medium is the message'.[3] In other words, the content of media is less important than the impact of each medium at social, psychological and sensory levels. McLuhan also introduced the vivid notion of the 'global village', which described the connection of the world by electronic media. He claimed that transmissions by satellite and other relays (Sputnik having circled the Earth in 1957 and turned it into the content of a mediated environment) had transformed society from the mechanical, objective, uninvolved and 'visual' world of print into an electronic one which was immersive, involved, immediate and 'acoustic' (see the Glossary at the end of this book). McLuhan spoke of a new 'tribal' electronic consciousness that would replace the individuated culture that had dominated the West since the invention of the printing press.

This essay explores the encounter of Marshall McLuhan's major insights into media and technology with the present world of information networks, e-commerce, digital technology and the age of virtual reality. The encounter was one which McLuhan did not live to witness, yet it has been staged under these conditions by those who find a new relevance to his ideas. My brief discussion examines some of the reasons for McLuhan's return, before analysing in more detail the relationship of his main speculations to the current period which I have called 'virtuality' – a broad term which will require some explanation.

Perhaps the most salient example of this liaison between new medium and old messenger is the journal *Wired*, which took the Canadian intellectual as its patron 'saint' of the brave new world of technology, art and communication in 1993. McLuhan's controversial insights were hotly debated and widely disseminated in the 1960s, but to a large extent they were considered marginal and superseded by the time of his death in 1980. However, they now seem appropriate for a medium that arrived just after his message had faded.

McLuhan's legacy for the new generation of academics, journalists and hackers is often phrased as a religious and prophetic one. Paul Levinson said: 'The handwriting for coming to terms with our digital age was on the wall of McLuhan's books.'[4] Others take a more analytical approach to the relationship between McLuhan's canonisation and the religious narratives of his writing. For example, Huyssen elegantly uncovers the theological programme running beneath the McLuhan interface, in which McLuhanist topics such as 'electricity', 'medium' and 'global village' can be replaced by 'Holy Spirit', 'God' and 'Rome' respectively.[5] In a similar manner, Genosko critically inquires whether McLuhan's 'fooling around' was founded on a specific faith, whereby salvation from the fall of literacy might be found in electric technology.[6]

Richard Coyne's philosophical study of new technology makes the link between spiritual values and a romantic sensibility towards our new media.[7] He includes in this scenario McLuhan's proposition that a translation of our entire lives into a spiritual form of information would transform the globe and the 'human family' into a single consciousness.[8] The philosophical and romantic aspects of his work will be explored below.

McLuhan's revival has its problems, and I wish to address these before exploring McLuhan's key thoughts in relation to virtual technology and geography. For example, he has been reprocessed for a new technology and to a certain extent has been sanitised by it, for his explorations are in some cases massaged to provide an affirmative, theoretical slant for the more utopian and idealist discourses of virtuality. As Ostrow remarks, there is no place in the recent, rhapsodic versions of virtual discourses for consideration of McLuhan's belated warning that the impact of the new electronic mass media could be harmful and might need monitoring.[9] Thus, McLuhanism in the new century tends to perpetuate the simplifications of his critics and supporters in the previous one.

It took the Situationist, Guy Debord, whose

writing on media and society set him in stark ideo-
logical contrast to McLuhan, to make the obvious
point that had been ignored by the 1990s
McLuhanites: McLuhan's optimism about the
potential for freedom and open access of the media
had in fact declined by the late 1970s. Debord had
identified what he called the 'society of the spectacle',
defined as a social relation among people, mediated
by images. Capitalistic domination had reached a
stage where it could alienate, subjugate and domin-
ate not just through the workplace, but through
commodity culture in all its variations, including
television and shopping. This theme would later
be developed in the work of Jean Baudrillard,
whose writing on 'simulation' – the production of
social reality through 'codes' such as the mass
media – took the concept of spectacular society to
its limits.

Debord noted that McLuhan had begun to recog-
nise the negative effects of the emerging corporate
and global media spectacle. In 1990, Debord even
wrote that McLuhan was abandoned by disciples
who were anxious to distance themselves from his
gloom, in order to get jobs in precisely these bloated
industries. For Debord, McLuhan, 'the spectacle's
first apologist, who seemed to be the most con-

vinced imbecile of the century, changed his mind when he finally discovered in 1976 that the pressure of the mass media leads to irrationality and that it was becoming urgent to modify their usage'.[10] At least Debord did McLuhan the service of noting his change of tone.

To flesh out the circumstances of McLuhan's improbable return, I should outline his decline in order to assess the defects in his work, prior to applying it to the new technological media.

Deleting McLuhan

The Canadian literature professor, Marshall McLuhan, who reluctantly turned his attention to the structures of advertising, television and the early electronic computer, had fallen out of favour by the late 1970s. One reason for this was his failure to combine his technological inquiry with a political one. He understood politics as simply a response to technology: democracy, for example, would break down – as would the electorate – in the decentralised instantaneity of electronic communication. His model of media did not accommodate the role of political activity. For example, his statement that Afro-American culture was primarily in tribal relationship to a mechanical world, and caught

8

between literacy and new electric media, could not really become an operative message for the diverse ethno-politicisation of academic discourse and its recognition of pluralist society. Genosko suggests that: 'The "backward is really superior" thesis was nothing less than a policy of repression completely lacking a political economic analysis of American racism and an acknowledgement, in any specific – embodied or historically situated – way whatsoever, of the history of slavery and the systematic eradication of Native Americans . . .'[11] McLuhan's primitivist and evolutionist views would not sustain critical analysis within recent discourses of ethnicity and 'post-colonial' studies.

McLuhan's deterministic and monolithic account of media necessarily foregoes a detailed analysis of political dynamics that shape and exploit media in different ways. Nothing that McLuhan said could adequately articulate the relationship between media, power and commerce. Jonathan Miller accused him of an 'abdication of political intelligence'.[12] Even his disciple Arthur Kroker tempered his own enthusiasm when he realised that McLuhan had no theory to analyse or interpret the relationship between the economy and technology or between corporate power and information, and substituted for political consciousness a contemplative stance of apolitical objectivity.[13]

Benjamin DeMott, in an observation that predates more recent criticisms of postmodern discourse's abandonment of reality in favour of surfaces, said that McLuhanism will make us 'rise to the certainty that style and method are all, that the visible – Vietnam or wherever – is not in any sense there. And having done this we can take off absolutely, fly up from the non-world of consciousness into the broad sanctuaries of ecstacy and hope.'[14] He quotes McLuhan's optimism about the computer, the technology of which promises a 'Pentecostal condition of universal understanding and unity . . . a perpetuity of collective harmony and peace'. DeMott prefers the term 'McLuhanacy' to describe this impossible world.

The sense of McLuhan's failure was acute for other critics who, in Levinson's view, were ignorant of the impact of his thought. He took umbrage with Bliss's review, which claimed that 'Once exalted as oracular, Marshall McLuhan's theories now seem laughably inadequate as an intellectual guide to our times'.[15]

By 1974 McLuhan had reiterated his message that media needed to be understood in terms of their radical effects, in order to wake their users from a 'self-induced subliminal trance'. Otherwise they would be 'slaves'.[16] However, this version of consciousness-raising is still phrased as a matter of the viewer becoming aware of the medium's means of circulating information and their impact on individual and society. But does this realisation amount to liberation? Once this transcendence is achieved, then what? Waking from a media-trance does not in itself dissolve the political economy of media.

European thought, which brought with it deconstructionism and post-structuralism, seemed further to seal McLuhan's work from current theory. As Genosko says, he had a derisive view of Derrida and French philosophy, even though French intellectuals had begun to respond to his own ideas. As we shall see, the lessons of deconstruction stand between McLuhan and his application to virtuality.

McLuhan's insights on television, radio and other technological extensions had certainly caught the imagination of the media and provoked debate in academia. McLuhan's declared intention was to place his culture under sustained observation in order to show how the future might reveal itself in the present. He insisted that he did not predict the future, but probed for its current effects. However, the relatively slow development in media technology meant

that his ideas could not be applied in altered circumstances. They therefore exposed themselves to the
allegation that they were *ex post facto*, revealing
what was already there, while being incapable of
being tested as theory on a new object. No other
technology had arrived to reveal whether his observations and views might be valid in the context of a
new media environment. The corollary is that
McLuhan has been criticised for proposing a paradigm for media that was insightful conjecture at
best and incoherent, generalised inquisition at worst.

McLuhan's popularity also declined because of
his rather ambivalent relation with corporate industry and media. By the mid-1970s, McLuhan's fame
had peaked, and through the rest of the decade his
popularity and influence waned. He had over-
exposed his work in the media and had become a
'personality' (even appearing in Woody Allen's
Annie Hall, 1977).[17] His co-authored books were
receiving few reviews, and a changing political
climate sidelined him. People 'found that he was too
opportunistic; they found that he was too easy to
buy, he was too available on the speech-giving
circuit, and somehow a bad aura collected around
him . . .'.[18] As Robert Fulford adds, McLuhan set
out to sell his ideas to business and governments for

money, and 'in order to do that he had to be famous; he wanted to be famous'.[19]

McLuhan's declining health understandably hindered his work's influence. He had been ill for some time, having initially had surgery for a brain tumour in 1967. He suffered a major stroke in 1979 and, cruelly, the man who favoured an extempore form of address was reduced to the use of a few words at a time. The man's cherished oral medium ('speech is the extension of thought') had become severely impaired.

In 1980, the Center for Culture and Technology was closed by the University of Toronto, either because McLuhan could not run it or, in Marchand's view, because it lacked funds. 'McLuhan showed up one evening and wept because his old office was in shambles.'[20]

McLuhan died in his sleep on 31 December 1980, on the very eve of the decade that heralded the greatest media revolution since Gutenberg. He was not able to witness the mass commercialisation of the computer, in which the microprocessor, the silicon chip, Bill Gates, Apple, Netscape and the modem inaugurated the age of the PC, the Internet, the World Wide Web and Virtual Reality. Neither could he engage with the accompanying discourses of cyberpunk, posthuman and cyborg theory and the explosion of academic courses in digital media. However, McLuhan's ideas did not disappear for long, and a combination of hard work by his inner circle and the realisation that his explorations might actually be relevant to the virtual age ensured his resurrection, aligning the foremost explorer of media with the new, virtual 'extensions of man'.

McLuhan's Cultural Economy

You can be quite sure that if there are going to be McLuhanites, I am not going to be one of them. I know that anyone who learns anything will learn it slightly askew. I can imagine that having disciples would become a very great bother.[21]

Marshall McLuhan

Before looking in detail at McLuhan's ideas and the problems of applying them to current media, I will outline the character of his revival. How has his work been reconstituted for the new century?

An answer lies in academia and the reinstitution-alisation of McLuhanism by former colleagues and students in Canada. Levinson and other writers on McLuhan have sought to maintain his standing and insights, and are therefore justifiably capitalising on the new technologies in order to reassert McLuhan's status. They are also testament to the new McLuhanite cultural economy which has benefited from the resurgence of the University of Toronto's McLuhan Program in Culture and Technology, after the place was shut down rather too precipitately following his death in 1980. It was reinvented as a 'McLuhan Studies Room', and launched along with an *Understanding McLuhan* CD-ROM. A significant number of books and websites of interest have been written or edited by ex-students, relatives and other participants in his courses or projects.

McLuhan's 'rediscovery', as his son Eric describes it, is assisted by the re-circulation of his ideas by his admirers,[22] yet Neo-McLuhanism could not be introduced to the culture unless prevailing techno-environmental conditions were suitable. The technological advances in computing and communications have enabled his supporters working in fields such as media studies to conjoin his work with the information revolution.

This task is, superficially, a simple one of referring to the shift in the technological world since his death. For example, McLuhan's son observes that the last twenty years have seen the arrival of new media. These include the personal computer, fax, video-conferencing, virtual reality and CD-ROMs. Further inventions have arrived, such as WAP (Wireless Application Protocol) phones and MP3 digital music.

However, technological change *per se* does not fully validate the reintroduction of McLuhanism as a tool for exploring new media and environments. The question of his relevance remains to be answered. The challenge is met using a number of strategies.

A spurious one is used by Eric McLuhan, who claims that nobody else is studying these new forms of media, and thus ignores the avalanche of academic and popular writing on these developments, some of which I discuss below.

Another is to flag up the critically reflective perspective of McLuhan's thought. For example, Moos emphasises the educational value of his message, claiming that he provides the means to develop an awareness of the social and other effects of media. This consciousness-raising amounts to a 'form of civil defense against media fallout', in which the central role of culture should elevate creative processes necessary for survival.[23] This accords with McLuhan's promotion of artistic activity as an early-warning system for society (which I discuss towards the end of this essay). It also connects with his criticism of educational deficits within the school, college and university systems, which failed to take critical account of the structural importance of electronic media, particularly television. This perspective highlights the capacity of McLuhan's thought to accommodate itself to future circumstances, particularly in his later, salutary writings on the potentially catastrophic effects of media – which include lines such as 'World War III is a secret dimension inherent to our own technology'.[24] Of course, Moos applies this *aperçu* to 21st-century technology in a conditional manner, in order to suggest that McLuhan *would* have drawn significance from the origin of the Internet in military communications technology. He also forces McLuhan's more esoteric ideas about computers into a scenario which presents the remote danger that artificial computer intelligence (see below) could lead to a 'type of World Wide Web that downloads WW3'.[25] McLuhan's work is phrased as a kind of pre-emptive detection device which, with a bit of an upgrade, will permit us to approach new technology with our understanding and perceptions attuned to the environment. As McLuhan said in

1967: 'If we understand the revolutionary transformations caused by new media, we can anticipate and control them.'[26]

McLuhanism is therefore cast as an educational tool which, if properly used, can expose the mechanisms by which media exert their effects on humans.

The return of McLuhan, however, necessarily raises questions about the specific relevance of his work to today's information networks. The interim between the decline of his influence and his reintroduction has seen both radical technological transformations and an emerging body of theory and study of their effects. McLuhan, therefore, is playing 'catch-up' in many respects. In order to do this, McLuhanism must test its broad assumptions against these new technological and intellectual environments, before the detail of its thought can be applied to the problematic concept of 'virtuality'.

Testing McLuhanism: The Problem of Prediction

He never predicted the future, never tried to. 'I'll tackle the really tough one: the present. Let me see if I can predict the present.'[27]

Eric McLuhan

[A future medium like a kind of computerised ESP would process] *consciousness as the corporate content of the environment – and eventually maybe even* [lead to] *a small portable computer, about the size of a hearing aid, that would process of* [sic] *private experience through the corporate experience, the way dreams do now.*[28]

Marshall McLuhan, 1967

It is assumed that McLuhan ignored the future in favour of an analysis of his present culture. He saw his role as dissecting the latter in order to supply humanity with the means to institute a heightened awareness of advances in technology. This had its historical dimension, because he strove to understand the television age in relation to long-term historical shifts and geographical differences in media and technology. His books had outlined the development and effects of transitions from speech, through alphabet and print, to electric media such as telegraphy and radio, in order to illuminate the cultural environment of his time.

Indeed, rather than phrasing the present in relation to the future, he emphatically argued that the present was an effect of the past. This is McLuhan's 'rear-view mirror' version of the media, which are initially understood and conceptualised in terms of previous technologies. 'When faced with a totally new situation', he explains, 'we tend always to attach ourselves to the objects, to the flavor of the most recent past. We look at the present through a rear-view mirror. We march backwards into the future.'[29] A typical example is the use by television of the previous content of film, or the employment by the car ('horseless carriage') of the visual language of a previous medium of transport. Computers use the technology of the typewriter and the contents of television, telephone, fax, newspaper, etc.

In other words, McLuhan attempted to provide theories based on historical evidence. However, while he sought to provide tools for understanding media that could be applied to future circumstances, he was reluctant to adopt the role of futurologist.

In terms of technological advances in media, this stance makes McLuhanism vulnerable to accusations that its power is blunted by its reluctance to get involved with detailed research on the political economy of media and its effect on media technology. His thought tends to suffer from a form of theoretical inertia, which renders his broad thesis too inflexible to account for the flux in the structural dynamics in media and communication.

For example, Castells claims that diversification of media (such as cable, digital and terrestrial TV) has led to a more extensive targeting of audiences according to lifestyles, and encountered a new dynamic of communication. This diversification and nicheing places McLuhanism under some duress. Castells argues that: 'While the audience received more and more diverse raw material from which to construct each person's image of the universe, the McLuhan Galaxy was a world of one-way communication, not of interaction . . . it fell short, McLuhan's genius notwithstanding, of expressing the culture in the information age. This is because processing goes far beyond one-way communication.'[30]

Castells contends that the first-generation utopian, communitarian and libertarian culture of users constructed the Net in two opposite directions. One was based on the restricted access to computer hobbyists with their pioneering spirit and distrust of commercialisation. The other was populated by 'newbies', for whom counter-cultural ideology still retains its informality and who subscribe to the principle that 'many communicate with many', and yet each person has a 'voice'. However, these newbies will reflect commercial interests because they will

21

extend the power of major public and private organisations into the realm of communication. Castells therefore concludes that while this group has its commercial dimension, it is still situated in the culture of individuality: 'Unlike the mass media of the McLuhan Galaxy, they have technologically and culturally embedded properties of interactivity and individualization.'[31]

In this context, even McLuhan's dictum, 'the medium is the message', is obsolete. For Castells, the message is the medium: the characteristics of the message shape the medium. For example, the medium of MTV is tailored in its entirety to its targeted youth audience.

At a more banal level, and despite McLuhan's claims to avoid prediction, his assumptions about technology also demand some considerable upgrading of his key ideas. In our age of networked computers, it is on face value prescient of McLuhan to use terms which are now familiar to us: 'The computer can be used to direct a network of global thermostats to pattern life in ways that will optimize human awareness. Already, it's technologically feasible to employ the computer to program societies in beneficial ways.'[32] Elsewhere he states that: 'You could run the world's biggest factory in a kitchen by a computer. With telephones, telexes, and computers – all of which operate at instant speeds – management and all forms of hardware can be centralized. The computer, literally, could run the world from a cottage.'[33]

McLuhan's observations reveal the limits of his thought, not only in relation to future technology but also in relation to the conclusions he drew from the media environment around him. His statement that electric technology leads to global instantaneity and decentralisation does not accord with his assumption that the technology of media still performs in centralist terms, whereby the world can be run from a localised database. His 1969 interview in *Playboy* demonstrates that his depiction of media as a homogeneous, one-way form of communication intrudes on his view of the computer. In the tradition of early science-fiction film, he suggests that the computer will become capable of programming media to 'determine the given messages a people should hear in terms of their over-all needs, creating a total media experience absorbed and patterned by all the senses'. Thus, McLuhan trades on a social-engineering view of computer technology, in line with the logic of the advertising industry. The mainframe will be capable of 'programming five hours less of TV in Italy to promote the reading of newspapers during an election', or provide a further 25 hours of TV in Venezuela to 'cool down the tribal temperature raised by radio the preceding month'. This ambitious prognosis is matched by McLuhan's fanciful idea that computers might be directly linked with private and collective consciousness and be able to process and communicate pure thought between the minds of individuals. These visions allow us to discern how the techno-romanticism of his time jars with the emerging techno-realism of our own, and blunts his predictive potential.

McLuhanism needs to be aligned with the dynamics of current computer-mediated communication, and attempts have been made to do just that. However, to explore this connection, we must first present the theoretical framework to which his ideas are being re-introduced, in order to inquire whether subsequent theoretical disciplines constitute obstacles to his new reception.

A Proto-Postmodernist? Theory Since McLuhan

A problem arises in the confrontation between McLuhan's model of media and subsequent theories and methodologies.

The issue, therefore, of McLuhan's relevance to our current situation must deal not just with technological advances in media, but also with discourses that have grown up around it, particularly the theories of 'virtuality' that have arisen since his death. In other words, McLuhanism must engage with theories that it either chose to sideline in its heyday or must now overcome in order to be viable.

A new development in the extensions of social communication had appeared which could put McLuhan's key ideas to the test, most notably Baudrillard's work on simulation. As Gary Genosko states, however, they had different agendas. Genosko argues that important theorists such as Paul Virilio and Jean Baudrillard extend and often implicitly criticise McLuhan's thought. Jean Baudrillard has been termed 'the French McLuhan', and his work in some respects bears superficial resemblances to the Canadian's views of media.[34] However, in the writing of both Baudrillard and Virilio a note of pessimism, if not cynicism, has replaced the positive tone that McLuhan set in the 1960s. For example, Virilio states that 'At the end of the century, there will not be much left of the expanse of a planet that is not only polluted but also shrunk, reduced to nothing, by the teletechnologies of general interactivity'.[35]

In respect of the foundations of postmodern thought, McLuhan's early work has been compared with that of other writers on popular culture, including Roland Barthes and structuralists such as Lévi-Strauss, who attend primarily to the structure of the medium (i.e. language) rather than its content. Inherent to these claims is the notion that McLuhan is a prototypical postmodernist whose explorations of media anticipate the postmodern perspective that is marked by its emphasis on social fragmentation, pluralism and the full emergence of a consumer culture dominated by simulations, or signs detached from referents.[36]

McLuhan's texts and conception of language arguably prefigure postmodern theories of the writer, reader and textuality. For some, McLuhan's style prefigures the virtual world and outdates its own medium. His books are often paratactic in style – meaning that they are constructed from a disconnected set of propositions. In some cases they use graphic and photographic means to disrupt the reader's assumptions about the linearity or 'logic' of reason, while providing a visual 'hit' that corresponds to the total, non-hierarchical text of newspapers rather than the expository 'ABD and E'ness' of a linear text. The relative novelty of his style ensured that it was occasionally criticised for its elliptical, non-sequential, non-academic structure.

In the light of new technologies, this version of writing is revealed as a prescient form that wanted to escape the confines of linear literature. For example, Levinson argues that some of McLuhan's books, such as *The Medium is the Massage* [*sic*], pre-dated the hypertext of the Web (e.g. the links on home pages) because they were constructed from a 'mosaic' set of fragments of text and image that could effectively be read in any order and in any direction according to the desires of the reader. It should be remembered that McLuhan was exploiting literary forms of montage and collage that are modernist in character, and his discussions do not stray far from references to typical examples such as James Joyce and Wyndham Lewis. Thus his relation to postmodernism must be qualified by the strong attraction he had for stylistic devices originating in early 20th-century literature.

The difference between McLuhan's work and high postmodern style and theory is not restricted to the issue of McLuhan's modernist references. It also extends to the philosophical assumptions on which his thought operates.

Postmodern culture is often broadly defined and

debated as a triumph of image over reality, surface over depth, style over content and signifier over signified (or referent). In this paradigm, the primacy of a text's meaning and possible interpretation is construed as an effect of a constant *semiurgy*: the mobilisation and reconfiguration of signs in endless combination with multiple effects. Postmodernism therefore focuses attention on language and its multiplicity of codes. The extreme post-structuralist position argues that the world *is* a text, and that reality is a discursive construct. This textualisation of the world has an impact on McLuhanism, as I shall briefly show. It is characterised in part by the work of Jacques Derrida, whose deconstructive strategy asks us to consider whether McLuhan's historical assumptions about the primacy of speech over print and television may be misplaced.

To demonstrate this, Derrida reveals the subordinate position that the text previously occupied in history to speech and the spoken word. It is in this context that Derrida's thought poses a 'deconstructive' question to McLuhan. Does McLuhan consider the relationship between *text* and *speech* in his own writing? Seemingly not, since McLuhan presupposes that speech (oral/acoustic culture) must precede writing (visual/linear culture).

We can illustrate this decisive contention between McLuhan and Derrida by comparing the opposed use they make of Plato's distrust of writing. McLuhan quotes the famous passage in Plato's *Phaedrus* in which the invention of writing is condemned:

'. . . *this discovery of yours will create forgetfulness in the learners' souls, because they will not use their memories; they will trust to the external written characters and not remember themselves.*'[37]

This supports McLuhan's version of events, because 'until literacy deprives language of [t]his multi-dimensional resonance, every word is a poetic world unto itself'.[38] Thus, in line with Plato, McLuhan privileges speech/word over text/alphabet.

We can now consider Derrida's reading of the same passage in Plato. A completely different position results from Derrida's deconstructive criticism of Plato's negative verdict on writing, and, in particular, the function of *metaphor* (the description of one thing by another) in language. Plato sees writing as poisonous to the primacy and full *presence* of speech. Metaphors draw philosophy away from the immediacy of speech, because they are inherently *at a distance* from what they describe. Derrida deftly signals that in fact Plato himself is exploiting metaphor – for example, writing as a 'poison' – in order to secure the desired presence of speech.[39] Plato's defence of speech attacks writing as harmful, and, at the same time, represses the fact that it relies on writing.

McLuhan's work can therefore be deconstructed in a similar manner to reveal that his triphasic model of evolution in media (speech, writing/print, electronic media) is based on the metaphysics of the presence or immediacy of speech and the relegation of writing to a subordinate position. Genosko has come to McLuhan's rescue, against Derrida's criticism. McLuhan's vision of the end of the book (writing/print), in Genosko's view, announces the beginning of television, which has an 'aurality' and tactility irreducible to speech.[40]

Genosko's attempt to circumvent the challenge of deconstruction does not fully account for McLuhan's insistence on the primacy of 'speech' presence. Arguably, in McLuhan's paradigm, even the medium of television relies on this same metaphysical ghost of presence. Indeed, to go further, we could say with Derrida that the virtual reality environment is itself crucially dependent on maintaining presence and the immediacy of speech (the virtual discourse insists on the claim that 'you are *really* there').

McLuhan's place in the postmodern constellation is therefore a contradictory one, in which his subject-matter appears to chime with more recent theories of culture that emphasise technological factors. While McLuhan trades in the themes that, superficially, define postmodernism (e.g. the medium replacing the meaning), his model is constructed on a search for origins and a transparency of communication – assumptions that postmodern theory attempts to debunk. His thought is locked into a mode that could not be identified by McLuhan unless he had absorbed the lessons of deconstruction. It is therefore simplistic to describe him as a postmodernist *avant la lettre*.

With these caveats in place, we can approach the central issue of this book: McLuhan's encounter with 'virtuality'. The first task is to define this term, or rather demonstrate the polysemic and contested character of this discourse.

Understanding Virtuality – Links Between McLuhan and Narratives of New Media

If we let slip a yawn at the mere mention of virtual reality, cyberspace, and embodied virtuality, or roll our eyes at the naming of telepresence, teletopia, and electronic cloning, it is because something has been missed in the headlong rush to exit the common-or-garden experience of everyday life for the apparent wonderment of the latest technologies.[41]

New technologies and media have not only been accompanied by discourses of virtuality, but have been constructed by them. They all have a bearing on the assumptions that accompany McLuhanism. The current discourse of virtuality springs from science, literature, philosophy and socio-cultural studies. The arrival of the computer has been presaged, dissected and debated by theory and literature, including science fiction. Cyberpunk, for example, is a genre that arose in the mid-1980s. It often represents a dystopian world of the near future that is organised and controlled along corporate capitalist lines and transformed by new technologies which alter the body, provide new forms of media, and construct the new geography of 'cyberspace'.[42]

The word 'cyberspace' was coined by cyberpunk writer William Gibson, yet it is not a definition confined to science fiction. Rather, it has become a literary narrative that has affected 'the way that virtual reality and cyberspace researchers are structuring their research agenda'.[43] The vision of cyberspace that Gibson offers dovetails with dominant versions of postmodern thought. For example, Fredric Jameson, David Harvey and Manuel Castells discuss 'global spatial totality', 'time-space compression', globalisation' and 'the space of flows'. Arguably, these models are sometimes over-generalised, and often too vague to provide analytical power.

However, these paradigms have recently come under more critical scrutiny. A history of the language of virtuality has thus grown out of a relatively uncritical and initially crude set of assumptions and become more measured and reflective. I wish to outline this development in order to show how certain narratives are organized according to certain philosophical presuppositions. These broad features of 'digital discourse' should allow us to introduce the new context of McLuhan's thought.

Michael Heim distinguishes between two approaches to the term 'virtual reality' introduced in 1984 by Jaron Lanier (the founder of VPL – Visual Programming Languages). Heim's typology is relevant because it demonstrates the current elasticity of the term, and also the extent to which culture confuses 'the artificial with the real, and the fabricated with the natural'.[44] Lanier patented the dataglove, the head-mounted display and the data-suit – with the purpose of integrating them into what in 1987 he called 'a reality-built-for-two' or 'RB2'.

The whole point of virtual reality (VR), according to Lanier, is 'to share imagination, to dwell in graphic and auditory worlds that are mutually expressive'.[45] For Heim, this constitutes the 'strong' technologically determined version of virtuality, where virtual reality is an emerging field of applied science. The strong meaning refers to a particular kind of technology rather than a consensual hallucination, simulated drug trip or illusion that the computer supplies us with. This version exhibits the typology of 'immersion', 'interaction' and information 'intensity'. The sense of immersion 'comes from devices that isolate the senses sufficiently to make a person feel transported to another place'.[46] 'Interaction' describes the computer's ability to change a virtual scene, in which the user is immersed, in synchronisation with the user's own movement and point-of-view. Information 'intensity' defines the degree to which the virtual world can offer users information about their environment. This can lead either to the VR characters displaying sentient behaviour (they behave like real entities) or to the experience of 'telepresence' – in other words, the extent to which a user feels present in a virtual environment. This could involve a link with another environment from a distance (e.g. controlling a real robot on Mars within a VR environment which sends a high degree of data from Mars, converted into a high intensity of information for the user).

'Virtual' has been applied not just to technology and not simply to the experience of interacting with simulations that have some computerised component. It extends into the 'real' itself. Heim states that this is the 'weak' or loose definition of VR. Everything – from automated teller machines which fulfil the function of a bank teller in a virtual (an 'as if') mode, to phone sex, e-mail and supposed 'real-life' experiences such as window-shopping – has now been dubbed 'virtual'. From its technological specificity to its use as a description or connotation of a 'condition', the meaning of virtuality has haemorrhaged.

McLuhan, of course, had no knowledge of virtual technology, so his version of virtuality is a 'weak' one, in which the virtual reality is expressed as the (ontological and epistemological) condition of 'as if', rather than being descriptive of a specific technology. However, his comments on the computer and on technology can still be related to immersion, interaction and information intensity in the loose sense. These themes are condensed in his theory of 'hot' and 'cool' media.

McLuhan says that 'A hot medium is one that extends one sense in "high definition". High definition is the state of being well filled with data.'[47] For example, a photograph is high definition (a high intensity of information, in Heim's term), while a cartoon has a low level of information. Likewise, the telephone is low definition because the ear is given only a restricted amount of information compared to radio. The corollary of this is consonant with Heim's 'interactivity', because so little is given in a cool medium that much has to be filled in by the listener/viewer/user: 'Hot media are, therefore, low in participation, and cool media are high in participation or completion by the audience.'[48]

Finally, the notion of 'immersion' applies in general form to McLuhan's observation that 'Electric media transport us instantly wherever we choose. When we are on the phone we don't just disappear down a hole, Alice in Wonderland style – we are there and they are here.' When we are on the phone, on the air or presumably online, we are in a sense absent from ourselves and with the other. In a 1978 essay in *New York Magazine*, McLuhan said that 'the sender is sent. The disembodied user extends to all those who are recipients of electric information'.[49] The principles that Heim outlines are therefore present in McLuhan's work in a rather fragmentary and nascent form. This demonstrates the flexibility of the 'weak' definition of virtuality, and reveals the properties of virtuality inherent to all electric and electronic media.

There is an important philosophical dimension to discourses of virtuality that underpins the dynamics of immersion and interaction. The accession to virtuality is analogous to the pathway from the individual and imperfect world to the unified and idealised world of cyberspace.

The accompanying themes of perfection and completion necessarily invite philosophical critiques of the language of virtuality, and Coyne employs the term 'techno-romanticism' in order to provide a template to discern how romantic, idealist and empirical traditions co-ordinate virtual narratives. The virtual narrative is not simply a consequence or by-product of recent cyber-theory or cyberpunk analysis. It draws on philosophical systems of thought that have origins in classical philosophy. Coyne, for example, explores the techno-romanticism of information technology, and the rationalist and empirical traditions upon which techno-discourse draws. He suggests that the human condition emerged through the romantic and rationalist-empirical discourses, 'between unity and fragmentation, transcendence and order, the ineffable, and the presumption of language'.[50] These discourses develop claims that we can transcend the embodied reality of our world towards unity, using the force of information technology.

A cursory glance at these philosophical strands reveals the obvious relevance of thought to virtual narratives. Plato separated the world into the realm of our senses (where appearances and things can deceive) and the intelligible realm of ideas – ideal and unchanging forms and immutable Good. Plotinus (205–270 AD) later deployed Plato's doctrine to claim that the soul strives to escape the material body and embrace the unity of the ideal real. The link with virtuality is expressed in the mode whereby digital narratives have absorbed the idealism of this Neoplatonic concept of *ecstasis*: the release of the soul from the body. In some virtual narratives, the soul is replaced by the mind – 'the means of *ecstasis* is immersion in an electronic data stream, and the realm of unity is cyberspace'.[51] This idealism is echoed by romantic idealist philosophers for whom the analytical, categorising, measuring powers of rationalism and empiricism distract the individual into studying particulars and conceal the perfection of unity. For Coyne, empiricism's emphasis on representation of space – its reduction and division – still leads to techno-romanticism: 'If computers allow us to model, mimic, and represent reality, then they indeed allow us to alter perceptual fields, challenge and distort reality, and create alternative realities. So rather than countering romanticism, empiricism provides the conditions for technoromantic narratives to promote the transcendent potential of computer space.'[52] McLuhan's version of technological humanism can be considered within this techno-romantic dimension:

By surpassing writing, we have regained our sensorial WHOLENESS, not on a national or cultural plane, but on a cosmic plane. We have evoked a super-civilized sub-primitive man.[53]

This statement also indicates that McLuhan's narrative of technology and media is not just premised on the romantic theme of unification or transcendence, but also includes a crucial element that is not normally present in virtual discourses. It distinguishes

McLuhan's work from the pervasive and fashionable themes of posthumanism and simulationism, for his unique perspective is premised on a 'myth of return', via technology, to a pre-literate social reality. He sets up electronic (and presumably virtual) technology's destiny as the ability to turn society into a unified collective. 'We now live in a *global* village', he announces, 'a simultaneous happening. We are back in acoustic space. We have begun again to structure the primordial feeling, the tribal emotions (from which a few centuries of literacy divorced us) of a culture that preceded the invention of writing and printing.'[54]

At this time in history, according to McLuhan, the senses of 'man' were in harmony and completeness, and thus man was at one with himself and his environment. The role of speech and listening was central, and thus these societies lived in what he called an 'acoustic' world, in which communication involved highly interactive exchanges in which thought and action were inseparable. This culture was tribal, engaged, practical and unitary. When literacy arrived, the relationship between man and environment was changed by the new medium. Writing emphasised the visual rather than the oral and acoustic: 'A goose quill put an end to talk, abolished mystery, gave us enclosed space and towns, brought roads and armies and bureaucracies.'[55] It enabled us to lay out our thoughts in linear order and to conquer space by transporting them on paper. The mechanisation of printing – McLuhan's Gutenberg Galaxy – turned history into classified data, and the transportable book brought the 'world of the dead into the space of the gentleman's library'. But print also isolated the reader and silenced her voice and discussion. Telegraph brought the entire world to the workman's breakfast table. Electronic media brings us unification, in a techno-romantic fashion. McLuhanism claims ours is a brand-new world of 'allatonceness' in which 'Time has ceased, "space" has vanished'.[56]

Unity, return and harmony are achieved by technological means, yet it is important to note the foundational concept on which McLuhan builds his technological humanism. It is the principle of sensory harmony, in which all senses have the potential of working in equitable relationship with one another when the capacity of media is optimised. When media deliver the requisite means to enable senses to work in accord, communication becomes inherently transparent, direct, full and immediate.

McLuhan's narrative presents electric technology

as the apogee of this principle, because it facilitates all the conditions that are required for sensory harmony, transparency and immediacy.

McLuhanism is predicated on foundational humanist assumptions that aspects of postmodernism will seek to refute in terms of deconstruction. This is the issue at stake, despite McLuhan's application to obvious postmodern issues of globalisation, information and society. One key postmodern concern is with theories of language and representation in their relation to reality, whichever way it is defined. In a postmodern world, reality is sometimes alleged to have receded or been replaced by 'hyperreality', a universe of images and codes that produce the real in their own terms. Language is constitutive of reality. The condition of virtuality is one that presupposes the same binary relationship between the virtual image (or medium) and reality. In postmodern terms, the primacy of the image over reality connects with the virtual narrative which contends that virtuality has affected reality in some way. Heim therefore argues that cyberspace is a tool for examining our sense of reality. However, this may take not only the examination of reality as itself a given, but also assume too much about the relation of the virtual to the real. A brief description of the significant versions of the relationship between virtual and real is necessary, if we are to place McLuhan within virtual discourse.

The articulation of virtuality with reality has several complexions, each of which attaches certain relative values to each term.

The first model constructs virtual reality as a 'false approximation' of reality (a degraded copy, simulation or too perfect a version of it); the second claims that it is a 'resolution' or 'hyperrealisation' of the real.[57] These positions are noticeable in narratives which assume that the immersion by the user in a virtual system is concomitant with a removal from the fullness and instability of real existence. For example, N. Katherine Hayles says: 'As we rush to explore the new vistas that cyberspace has made available for colonization, let us remember the fragility of a material world that cannot be replaced.'[58] The second version assumes that reality seems impoverished and that virtuality can complete it, rather like glasses can 'resolve' poor eyesight. Reality has within itself the possibility of its supplement *and* completion in virtuality (called 'suppletion' of the real).[59] Thus, different versions of the real and virtual are presented. A third, more extreme version depicts virtual reality as a total resolution of the real, in which humans could escape from the world and into technology (as in Moravec's downloading of consciousness into computers):[60] we would be obliged by scientists 'to step out of the world without leaving a trace. We would never have been (t)here'.[61]

A further dimension to these views is outlined in Heim's analyses of the function of realism and idealism in virtual narratives, which he performs in order to construct a more pragmatic approach ('virtual realism') to the issue of virtuality than that achieved by utopian and dystopian writings on cyberspace from the 1980s. He argues that 'the network idealist builds collective bee-hives. The idealist sees the next century as an enormous communitarian buzz. The free circulation of information runs through the "planetary nervous system"'. Heim comments that for the network idealist, 'The prospect seems so exciting that you see the phrase "virtual communities" mentioned in the same breath with McLuhan's "global village" or Teilhard's[62] "Omega Point"'.[63]

McLuhan's view of a global community is often taken only in its positive, idealised state. However, Heim and others should be cognisant of McLuhan's qualifications to his thesis. McLuhan claims that: 'The global-village conditions being forged by the electric technology stimulate more discontinuity and diversity and division than the old mechanical, standardized society; in fact, the global village makes maximum disagreement and creative dialog inevitable.'[64] Problems of discourses on tribalism aside, the issue of the destructive effects of mass media and virtuality are often ignored in assessments of McLuhan's work. There is a danger of overemphasising the unitary, transparent and non-dialectical aspects of his thesis on the global village. Contradiction and conflict do exist in this equation.

The other side of the idealist coin is what Heim calls 'naïve realism'. This view of virtuality is one that defines virtual reality as a suppression of reality, to the point where computer systems are seen as alien intruders, where new media 'infiltrate and distort non-mediated experience until immediate experience is compromised'.[65] Attendant on this theme is the fear of losing local identity, interdependence and community as we merge in a virtual and global network. Naïve realism is not applicable to McLuhan's techno-idealism.

Finally, the fourth, radical and particularly postmodern account of virtuality contends that there is no distinction between virtuality and reality. This is because the real has always been virtual, as it is never fully present or actual. In other words, reality never intersects with itself, as it is constructed (like our identities) through difference and not presence. Language is the most obvious tool that separates us from ourselves while defining us to ourselves. This particularly 'weak' or loose definition of virtuality makes sense when the real is construed as socially and culturally constructed. For example, sociological method shows that our lives are always constructed through mediations and interactions.[66]

This dimension is paralleled in another form by psychoanalytic linguistic readings of reality and virtuality, or in the Lacanian trinity of the real, imaginary and symbolic. According to Žižek, if virtual reality is all surface, with no actual access to substance and the real, then we find this is also the case in real life outside virtual reality. Like VR, the real is surface, and it does not permit access to the 'true real'. Žižek notes that virtual reality shows us the virtualisation of the true reality: 'by the mirage of "virtual reality", the "true" reality itself is posited as a semblance of itself, as a pure symbolic edifice. That fact that "a computer doesn't think" means that the price for our access to "reality" is that *something must remain unthought*.'[67] For Coyne, this means that the ambitions of VR remind us that the real resists representation. In a romantic and sublime register, 'It is ineffable'.[68]

As we have seen, McLuhanism thrives on a triple narrative of initial unity (in primitive, oral cultures), fragmentation (in writing and print) and reunification (in electronic media). It provides an historical and spatial dimension, in which the proximity of 'man' to reality is contingent upon the technologies of media at each period in history or in each global location. Superficially, this might lead one to suppose that, for McLuhan, the relation between media and reality is conditional upon historical circumstances. However, McLuhan bases his analyses on a residual and ahistorical psychology of perception, wherein language/media/technology are oriented at all times towards the ratios between human senses: 'media are artificial extensions of sensory existence'.[69] Furthermore, for McLuhan, the optimisation of these senses – the parity and harmonious unity of sight, touch, sound – is the desirable objective of media. The electronic media enable this 'allatonceness'. He assumes that this unification and instantaneity, which defines new technology,

augurs a *return* to reality. His reality, then, is one
that is primarily socio-psychological but technologically enabled. In this framework, sociality has
historically been lost because media have not just
extended human senses beyond their immediate and
unified domain, but divided human senses and frag-
mented human identity. New electronic media now
permit a return to this lost reality, although
McLuhan warns that we must be careful to use
them correctly, and be assiduous in discerning their
effects. Thus, we return with a heightened critical
awareness to a state of collective, tribal consciousness. The return to social reality is analogous to a
pre-linguistic state, and McLuhan stipulates that
'the new media [e.g. television] are not bridges
between man and nature: they are nature'.[70]

In the context of the four versions of virtual-real
relations outlined above, McLuhanism therefore
requires some analysis of the definition of reality
that is presupposed in its discourse. McLuhan did
not involve himself with contemporary postmodern
and deconstructive approaches to the question of
the construction of the binary of language/media/
image and reality. He took for granted the narrative of technology and media as returning us to a
contradictory state of an advanced primitivism, but

did not question the assumptions that construed this social reality and its conditions. If we were to compare his presuppositions about the relationship between the media and reality with the discourses of the virtual-real, we might then describe McLuhan's project.

First, McLuhan would not see virtual reality as a false version of reality. As a medium – and thus an extension of man's sensory faculties – it could not be false in itself, for it is defined as a prosthetic or extension, in which such epistemological criteria are irrelevant. To define cyberspace as 'unreal' or inauthentic in comparison to reality is itself deceptive. Indeed, McLuhan did not provide a distinction between authentic and inauthentic, for media generate the perception of change in the first place, and are not merely false representations of a coherent reality. The question of cyberspace's failure to live up to reality would instead have to be assessed according to the extent to which it orchestrated and mobilised man's senses as a unified sensorium. It would be found to fail 'reality' if it perpetuated the disequilibrium and fragmentation of the senses.

Second, the reverse contention that reality is impoverished and that virtual technology leads to its completion is only applicable to McLuhan's thesis if the historical narrative of his work is forgotten. In other words, McLuhan is adamant that at certain moments in history 'reality' was not impoverished. Tribal communities had a fully functional and integrated form of communication which electronic technologies would re-install.

Third, the assertion that virtuality will lead to a full-fledged escape from reality can only be considered in the context of McLuhan's religious and romantic claims of return and unification, and relative historical and technological conditions. His reality is defined as existing prior to the historical development of media since print, but emerging again under electronic media. In this sense, virtuality will mean a return to, and not an escape from, reality – even in the case of McLuhan's more extreme ideas of the computer's potential to enable the communication of pure thought without a medium (thus rendering the empirical reliance on the senses unnecessary).

Finally, McLuhanism is ambiguous on the postmodern 'reversal' in which reality is construed as a textual, symbolic and absent construction rather than the immediate existence and 'givenness' to experience that common sense and empiricism assume is foundational, even if open to debate. On the one hand, the phrase 'the medium is the message' seems to shift emphasis from the world that media purportedly represent and reorganise (the 'content'), to the operation of the codes by which such content is dispersed in the environment. On the other hand, McLuhanism is only superficially an aspect of postmodern logic. This is because the modern themes of anthropological and psychological origination and return (back to primitive, unified sensory states), and the primacy accorded to speech and presence over writing and absence (Derridean deconstruction), are in contradiction to the postmodern paradigm. This distinction is made clear by Coyne: 'Thus when critics of electronic media argue that the new symbolic environment does not represent "reality", they implicitly refer to an absurdly primitive notion of "uncoded" real experience that never existed.'[71]

McLuhanism's encounter with virtuality is therefore a problematic one, because it can be reduced to the themes of virtual discourse, yet it demonstrates a blindspot to the central strategies of postmodern thought. This lacuna in McLuhan's work is understandable, given the trajectory of his thought from modernist examples such as T.S. Eliot. Furthermore, while we can confer upon McLuhan's principles the gloss of postmodern theory, we have to consider that he was working within fairly restricted parameters.

Finally, the encounter of McLuhan with virtuality has been of the 'weak' variety. Postmodern theory claims that there is no separation between reality and symbolic constructions: the world is based on the production and consumption of signs. How, then, do we separate this general view of reality as virtual, from the 'strong' account of real virtuality? In contrast to earlier historical experience, what sort of communication system generates real virtuality? An answer is that it is a system in which reality itself (which is to assume rather too much about reality) is captured, and in which appearances are not mediated by the screen in our experience, but *become* that experience.

This inherent technological aspect of virtuality – the strong version of virtual reality – now needs comparison with the specific probes which McLuhan used.

McLuhan's Probes

The next medium, whatever it is – it may be the extension of consciousness – will include television as its content, not as its environment, and will transform television into an art form.[72]

Marshall McLuhan, 1967

Marshall McLuhan's explorations of media are now being reassessed for the virtual age. The primary theoretical assumptions of his work – the rear-view mirror, visual versus acoustic cultures, hot and cool media, and the tetradic theory of media (see Glossary) – can be applied to virtual technology and virtuality, and tested accordingly.

McLuhan claims that culture works like a rear-view mirror, because new media render previous ones obsolete while taking them as their content (e.g. the TV takes film as its material). Levinson supports this view, arguing that the digital environment absorbs the early mass-electronic environment while enhancing the global reach of the latter.[73] However, Levinson avoids falling into a simplistic view of McLuhanism, which relegates literate culture to a subordinate position in relation to the immediacy and 'wholeness' of oral culture. He does this by making salient both the superimposition and synthesis of written, visual and oral media within virtual cultures such as the World Wide Web, and the aspects of McLuhan's work that underline the hybrid character of media.

McLuhan defines the hybridisation of media as a 'civil war', in which 'The crossings or hybridizations of the media release great new force and energy as by fission or fusion'.[74] These dynamic exchanges are most turbulent when they can be expressed at a broad cultural level, in the confrontation between literate and oral cultures. At this macro-level, McLuhan tends to align these cultural conjunctions with debatable constructions of nationality and ethnicity. The Chinese, for example, release explosive energy on the arrival of literacy; conversely, the West is destabilised by the oral and tribal 'ear-culture' that the electronic media introduces to its formerly literate, individualised, text-based culture.

Levinson combines the hybrid model of media with McLuhan's principle that a medium has as its content a previous medium. With virtual technologies, the upshot is that the medium which serves as content for the Web is not one medium, but is composed of many media. It is a meta-medium. Levinson proposes that the Web has taken as its content 'the written word in forms ranging from love letters to newspapers, plus telephone, radio ("RealAudio" on the Web), and moving images with sound which can be considered a version of television'.[75] Virtual technologies, in the loose sense, are able to function 'as if' they were other media, because the computer – the technology that enables virtual culture – can absorb and emulate all other media. Indeed, given the amorphous and chameleon character of virtual media, there may be a case for jettisoning the term 'medium' altogether when considering the 'matrix' that includes the Web and Internet. Perhaps 'medium' is too loaded and isolationist a description in the virtual world.

Levinson is therefore claiming that the written word is not obsolete, but has become transfigured in the new medium. For him, the interactive medium, in which users in virtual space operate in immediate 'real-time' by using speech, writing and images, is the combination that 'conspired to make online communication a speech-like medium, a hybrid in which our fingers not only do the walking but the talking, from its inception'.[76] The two-way model of interaction supersedes McLuhan's historically restricted version of 'acoustic' media. He assumed that a tribal, immediate and connected culture could only be discerned within electronic media on account of the potential for satellite technology to connect people instantly with global events. In the virtual culture, the instantaneity of electronic messages and collective response is nothing compared to the realities of immediate intercommunication on e-mail, through e-conferencing and via WAP technology. Importantly, this virtual community turns consumers into *producers* of their own texts and images. Interaction involves pro-activity. The postmodern view of the reader-as-writer has an impact on McLuhan's view of media. It propels his version of the human subject, whose senses are involved to a greater or lesser degree with each medium (the cooler the medium, the more participation or filling-in by the senses of the viewer), to new heights.

The digital era has therefore problematised McLuhan's definitions of hot and cool media. In his world and method, television was a cool medium, which paradoxically stimulated participation on account of its low definition and lack of intensity, yet it prohibited any direct interaction, resulting in the frustration of viewers who wanted 'to reach out and touch someone, to get in touch'.[77] However, virtual technology, at least in its interactive setting, resolves this lack and fulfils the desire to participate. McLuhan acknowledged that the reader of texts could become an active participant in their construction (his montaged, 'mosaic' style was designed to exemplify this 'writerly' approach to texts). However, he could not have countenanced

the extent to which the implosion of media consumption and production within the applications of
the PC and computer would alter the definition of
the user. In other words, despite his remarkable
ability to draw our attention to the use of electronic
media for linking users globally, he tended to under-
play the extent to which users would actively participate in technology and media. His assumption
about the degree of participation was framed by his
emphasis on oral, tactile and acoustic involvement,
in which a cool medium such as the telephone
demands active involvement. However, this view of
interaction presumes that users are participating
through a medium, and that the degree to which
they interact relies on the extent to which their
senses (such as hearing) are involved. With virtuality,
in its widest sense, the use of e-mail, e-conferencing
and other tools demonstrates the shift from
McLuhan's definition of the user as participant
through a medium to manipulator *of* that medium.
The telephone could not be manipulated as a
medium; however, the computer is much more
'plastic' in its interactive possibilities. A typical
computer monitor might have e-mail, Web chat,
RealAudio, word-processing and a news ticker
functioning at the same time. This would constitute

a broad spectrum of hot and cool media on one screen, and the onus for their combination would fall on the hot and cool attitudes of the user, who is constructing a media environment for herself.

Despite this relative blind-spot, the possibility of shifting the emphasis of McLuhan's dictum 'the medium is the message' to a less well-known version of his project is therefore possible, because he also suggests that 'in all media the user is the content'.[78] Indeed, the term 'user' has replaced those of 'viewer' or 'audience' in Internet culture. The ideology of involvement cuts to the core of virtuality, and the user becomes the medium through which the Internet operates.

Levinson's revision of McLuhanism does not account for all virtual scenarios. It is certainly the case that, in virtual terms, print has become less visual and more acoustic, because it is now subject to the laws of oral culture. It can be used to communicate with another person or people directly, in real time, without the disadvantages of traditional print media. It can be combined with other media on the Internet, and it can undermine the linearity of the book by exploiting hyperlinks. The ability to click on highlighted words in a text on the Web places the onus on the user to construct that text while reading it, demolishing the primacy of the author and the integrity of the book. This holds true, in theory, for other media that the computer has absorbed.

There are virtual contexts that McLuhan could not have envisaged. For example, the narrative of totality, in which users become part of a collective, does not operate in situations where they are immersed in a virtual world that has no demonstrable application to the global village. Computer games, for example, can be played in isolation, where the only contact is between a user and a program that has been constructed and marketed by a corporate industry. Virtuality invites the immersion of the user in an environment that has nothing to do with participation. However, this narcissistic feedback loop is buckling under the 'urge to merge', because most popular games are now networked. They enable businesswomen in New York to adopt on-screen characters in *Quake II*, *Doom VII* and other corporate games, in order to wreak virtual and reciprocated carnage on individuals from Princeton to Poole. Thus, McLuhan's repeated warning that the non-critical fascination of users with their medium – the Narcissus narcosis – would blind society to the benefits and disadvantages of its media, becomes the logic of virtual immersion in these cases.

Perhaps the most significant application of McLuhan's ideas to virtuality lies in his work on the tetrad (see Glossary), which was published posthumously. The tetrad was the culmination of his attempt to formulate laws of media in order to provide a scientific basis to his explorations, and to condense his disparate insights into media in one neat set of questions. The application of the tetrad to virtuality arguably presumes that, first, a computer can be taken as representative of the virtual condition in general, and second, that it can be considered a medium in the same way as, say, television.

By applying McLuhan's four laws or effects of media, we can answer the following questions.

To the question of what aspects of society the new medium enhances or amplifies, we can answer that it increases the participation in the medium that was merely suggested in the relatively low definition of television, while at the same time promoting interactivity between participants at a global level. In response to the question of the aspects of media dominant before the arrival of the medium that it renders obsolete, we can suggest that it partially eclipses the telephone, the typewriter, the paintbrush, the paper fax and the CD, to name a few. In answer to the question of what the medium returns to prominence from previous obsolescence, we can reply that it reinvents the written letter in the form of e-mail. Lastly, we can answer the question of what the medium reverses or turns back into when it has run its course or been developed to its fullest potential as follows: it is already turning into the WAP (Wireless Application Protocol) mobile phone, which now carries websites and e-mail. These portable, palm-sized devices will soon be able to show Keanu Reeves films and provide live 'narrowcasts' of Manchester City's football matches. The computer will be miniaturised and absorbed in the digital watch, and it will become the content of digital television. Eventually the desktop computer will be perceived as a hindrance and a quaint, primitive artefact.

The question remains, however, of whether McLuhan's ideas can transcend their own historical limitations, cast off their notions of global consciousness, 'dropping out', 'acting cool', *and* compete with prevailing and persuasive analyses of virtual culture. The effort to relate his probes to the present era of communications seems, in my view at least, hamstrung by a structural problem with his original work. This is its lack of engagement with the political economy of mass media, and its refusal to consider the content of media in any way other than as an irrelevance. There is no room in his thesis for analysis of the role of resistance to the message of the medium, and a cynic might argue that there is only a slim possibility that McLuhanism can survive without undergoing what amounts to a radical reorientation to socio-economic and political factors. A thorough critique of McLuhan's work would have to broach issues of multi-corporate global capital, access to new technologies, surveillance and censorship, monopolisation of software. Levinson has attempted to do this in his work on McLuhan, but the attempt to upgrade him in this context does seem restricted by the models that McLuhan prescribed.

This is not to say that the brave new world of virtual discourse is itself without its blunt edges. My final discussion addresses the dominant discourses on virtual identity and embodiment in relation to the work of McLuhan. Perhaps McLuhan can provide the means to temper more strident versions of cyborg theory.

Disincarnate Humans and Disconnected Identities

In the sense that these media are extensions of ourselves – of man – then my interest in them is utterly humanistic.[79]

Marshall McLuhan

McLuhan intersects with current debates on 'virtual identity' and the relationship of the body, identity and electronic technologies.

The discourse on virtual identity has three main inflections. Donna Haraway's cyborg theory, for example, underlines the narratives of suture, rupture and loss of bounded identity and their political potential. Her ground-breaking and provocative feminist manifesto for cyborgs proposes that such breaches of identity should be welcomed: the 'transgenic hybrids' that genetic engineering could create open up the possibility of a cybernetic, postfeminist libertarianism. The posthuman figure of the cyborg breaches the boundary between nature and artifice, body and machine. Haraway claims that this provides the entirely new alternatives of hybridisation to gendered identity being coded as 'natural' or 'artificial'.[80]

Conversely, Sherrie Turkle affirms that immersion in a networked computer can provide the user with a number of identities, so that the self becomes endlessly multiplied. While this actualisation of the self seems to echo Haraway's view, Turkle's construction of virtual identity does not lead towards radical disruption, contradiction and fragmentation, but to a condition in which the persona and the self converge. She points to the ability of the Internet to cultivate 'psychological well-being', and speaks of new identities as 'multiple yet coherent' in a permeable relation between the virtual and the real, 'each having the potential for enriching and expanding the other'.[81]

A third position draws attention to the relation of the body to new technology. N. Katherine Hayles asserts that 'At the end of the twentieth century, it is evidently still necessary to insist on the obvious: we are embodied creatures'.[82] This formulation is intended to counteract virtual theories that place too much weight on the connection between consciousness and the computer. It reintroduces the body into the equation, and thus provides a caveat against discourses that conflate the real with the virtual, in which the mind is unproblematically represented as escaping from the body into cyberspace.

To some extent, this position also correlates with Heim's plea for a pragmatic and cautious approach ('virtual realism') to the subject of identity. Heim asks us to acknowledge a complex relationship with computers, and avoid 'glib exaggerations such as "Now we're cyborgs", or "Everything's virtual reality"'.[83] Of course, this assertion could lead one to suggest that perhaps Hayles's emphasis on embodiment is itself an exaggeration and a generalised abstraction of the role of the body in the virtual equation.

McLuhan's ideas can usefully be plotted according to these discourses on virtual identity. For example, the issue of embodiment and disembodiment in virtual identity is echoed in McLuhan's image of 'disincarnate man'. However, this figure is removed from the ego psychology of Turkle, for its disembodied character is registered within the 'anxious' discourse of virtuality. Here, any positive effects of the sensory shifts that television induces have resulted in the disincarnation of the viewer – a consequence of a psychic shock that the individual undergoes when exposed to media that weaken his or her sense of having a physical body and an autonomous identity. This is particularly burdensome for younger members of society, for McLuhan claims that 'From Tokyo to Paris to Columbia, youth mindlessly acts out its identity question in the theater of the streets, searching not for goals but for roles, striving for an identity that eludes them'.[84]

McLuhan's version of disembodiment therefore cuts across superficial readings of the collectivism of the global village and the elevation of humans to a cosmic consciousness. For McLuhan, immersion in electronic media does not merely imply an elevation to a sublime state of global union, because his model incorporates the (admittedly under-theorised) conception that such immersion has a psychological and sensory impact that profoundly affects the ontological security of the individual. Presumably, virtual technology in its strict sense – virtuality as a condition – would exacerbate this threatening disincarnation or disembodiment: 'Mental breakdown of varying degrees is the very common result of uprooting and inundation with new information and endless patterns of informa-tion.'[85] Always the educationalist, McLuhan warns us that unless we are aware of this dynamic, we shall enter a phase of 'panic terrors, exactly befit-ting a small world of tribal drums, total interdepen-dence, and superimposed co-existence'.[86]

Conclusion: Virtual McLuhan

McLuhan's message that we should be made aware of the media, and send out constant probes in order to test their effects, constructs an important role for a critically reflective approach to virtuality. This mantle is conferred upon specific members of society, particularly artists and writers. McLuhan, following Ezra Pound, had stated that artists perform the role of 'antennae', using perceptions that are attuned to shifts in media, and thus behave as early warning systems. They therefore reverse the 'rear-view mirror' scenario, providing navigational guides for the new (virtual) environment, rather than making the old environment their content. This is a modernist avantgarde position, and perhaps it is presumptuous of McLuhan to define art as a process through which sensory awareness expresses itself in its most acute form. However, McLuhan's overarching maxim – that an awareness of the medium as a message is to be considered ahead of its content – is intended to provide critical tools that will enable others to perform McLuhanite probes into virtual media. McLuhan expressed the view that there would be no McLuhanites, in the sense that his probes were for him alone. This leaves the ground open to others who wish to take up where he prematurely left off. It is understandable, therefore, that the McLuhan facility at the University of Toronto has introduced The Virtual Reality Artists' Access Program, providing a virtual space to explore critically artistic uses of interactivity and transinter-activity: the 'dialogue of bodies interacting in a virtual tactile space'.[87] Perhaps this is a robust response to McLuhan's urgent announcement that 'The artist today might well inquire whether he has time to make a space to meet the spaces that he will meet'.[88]

The question of whether McLuhan's ideas them-selves will be transfigured to illuminate virtuality and its discourses remains intriguing. The Canadian satellite of McLuhanism continues to beam his message to the world, but it can no longer trade on the force of his personality, the magnetism of his lecturing style and his omnipresence in the press and on television. When McLuhan died, his influ-ence dissipated. Reviving it will require not only the virtual conferences, websites and embodied prac-tices of virtual art, but also the critical acceptance of the value of his work in institutes and on courses beyond the localised McLuhanite environment of Toronto. It remains to be seen whether the new technologies will be able to disseminate his message as effectively as Marshall McLuhan did through his chosen media.

Notes

1. Paul Benedetti and Nancy DeHard (eds), *Forward Through the Rearview Mirror: Reflections on and by Marshall McLuhan*, Ontario: Prentice Hall Canada Inc., 1997, p. 171.

2. Marshall McLuhan and G.E. Stearn, 'Even Hercules had to Clean the Augean Stables but Once!: A Dialogue', in *McLuhan: Hot & Cool* (ed. G.E. Stearn), Harmondsworth: Penguin, 1968, p. 302.

3. Marshall McLuhan, *Understanding Media: The Extensions of Man*, London: Ark, 1987, p. 7.

4. Paul Levinson, *Digital McLuhan: A Guide to the Information Millennium,* London: Routledge, 1999, p. 2.

5. Andrea Huyssen, 'In the Shadows of McLuhan: Jean Baudrillard's Theory of Simulation', *Assemblage* 10, pp. 7–17.

6. Gary Genosko, *McLuhan and Baudrillard: The Masters of Implosion*, London and New York: Routledge, 1999, p. 13.

7. Richard Coyne, *Technoromanticism: Digital narrative, Holism and the Romance of the Real*, Cambridge, Mass.: MIT, 1999, p. 304.

8. Ibid., p. 63. Coyne cites Wasson's claim that McLuhan reads the new technological environment as a 'book of symbols which reveals the Incarnation. Because everything in the world is a symbol, McLuhan can offer symbolic interpretation'. See R. Wasson, 'Marshall McLuhan and the Politics of

Modernism', *Massachusetts Review* 13 (4), pp. 567–80. See also McLuhan, op. cit., 1987, p. 61.

9. Michel A. Moos (ed.), *Media Research: Technology, Art, Communication: Essays by Marshall McLuhan*, Amsterdam: G+B Arts International, 1997, p. xvi.

10. Guy Debord, *Comments on the Society of the Spectacle*, London: Verso, 1990, pp. 33–4.

11. Genosko, op. cit., p. 108.

12. Jonathan Miller, *Marshall McLuhan*, New York: Viking, 1971, p. 76.

13. Genosko, op. cit., p. 79.

14. Benjamin DeMott, 'Against McLuhan', in *McLuhan: Hot & Cool*, pp. 282–3.

15. Levinson, op. cit., p. 29.

16. Benedetti and DeHard, op. cit., 'Making Contact with Marshall McLuhan', interview by Louis Forsdale, 1974, p. 198.

17. While he queues to watch a film, Woody Allen's character chastises a media lecturer for loudly pontificating on McLuhan's ideas and mistaking television for a hot medium. Allen happens to have McLuhan at the scene. 'You know nothing of my work', says McLuhan to the academic. 'You mean my whole fallacy [*sic*] is wrong. How you ever got to teach a course in anything is totally amazing.'

18. Benedetti and DeHard, op. cit., 1997, p. 174.

19. Ibid.

20. Ibid.

21. Marshall McLuhan and G.E. Stearn, 'A Dialogue', in

McLuhan: Hot & Cool, p. 335.

22. Benedetti and DeHard, op. cit., 1997, p. 183.

23. Moos, op. cit., 1997, p. 166.

24. Marshall McLuhan and Barrington Nevitt, *Take Today: The Executive as Dropout*, New York: Harcourt Brace Jovanovich, 1972, in Moos, op. cit., p. 165.

25. Moos, op. cit., p. 166.

26. Benedetti and DeHard, op. cit., 1997, p. 198.

27. Ibid., p. 186.

28. Eric McLuhan and Frank Zingrone (eds), *Essential McLuhan*, London: Routledge, 1997, p. 297.

29. Marshall McLuhan and Quentin Fiore, *The Medium is the Massage: An Inventory of Effects*, Harmondsworth: Penguin, 1967, pp. 74–5.

30. Manuel Castells, *The Information Age: Economy, Society and Culture; Volume I: The Rise of the Network Society*, Malden, Mass. and Oxford: Blackwell, 1996, p. 341.

31. Ibid., p. 358.

32. *Playboy* interview (1969), 'Marshall McLuhan – A Candid Conversation with the High Priest of Popcult and Metaphysician of Media', in E. McLuhan and Zingrone, op. cit., p. 263.

33. Benedetti and DeHard, op. cit., p. 177.

34. See Jean Baudrillard, *In the Shadow of the Silent Majorities*, New York: Semiotext(e), 1983, and *The Ecstasy of Communication*, New York: Semiotext(e), 1987.

35. Paul Virilio, *Open Sky*, London: Verso, 1997, p. 21.

36. David Harvey, *The Condition of Postmodernity: An Enquiry into the Origins of Cultural Change*, Oxford and Cambridge, Mass.: Basil Blackwell, 1989, p. 289.

37. Marshall McLuhan, *The Gutenberg Galaxy*, London: Routledge and Kegan Paul, 1962, p. 25.

38. Ibid.

39. Jacques Derrida, 'Plato's Pharmacy', in *Dissemination*, London: Athlone Press, 1981, pp. 61–171.

40. Genosko, op. cit., pp. 41–2.

41. Marcus A. Doel and David B. Clarke, 'Virtual Worlds: Simulation, Suppletion, S(ed)uction and Simulacra', in Mike Crang, Phil Crang and John May (eds), *Virtual Geographies: Bodies, Space and Relations*, London and New York: Routledge, 1999, p. 261.

42. James Kneale, 'The Virtual Realities of Technology and Fiction: Reading William Gibson's Cyberspace', in Crang, Crang and May (eds), op. cit., p. 29.

43. David Tomas, 'Old Rituals for a New Space: Rites of Passage and William Gibson's Cultural Model of Cyberspace', in M. Benedikt (ed.), *Cyberspace: First Steps*, Cambridge, Mass.: MIT Press, 1991, p. 46.

44. Michael Heim, *Virtual Realism*, Oxford and New York: Oxford University Press, 1998, p. 4.

45. Ibid., p. 16.

46. Ibid., pp. 6–7.

47. McLuhan, op. cit., 1987, p. 22.

48. Ibid., p. 23.

49. Levinson, op. cit., p. 39.

50. Coyne, op. cit., p. 7.

51. Ibid., p. 10.

52. Ibid., p. 106.

53. Marshall McLuhan, *Counter-blast*, London: Rapp and Whiting, 1970, p. 16.

54. McLuhan and Fiore, op. cit., p. 63.

55. Ibid., pp. 14–16.

56. Ibid., p. 63.

57. Doel and Clarke, op. cit., p. 261.

58. N. Katherine Hayles, *How We Became Posthuman: Virtual Bodies in Cybernetics, Literature, and Informatics,* Chicago and London: University of Chicago Press, 1999, p. 49.

59. Doel and Clarke, op. cit., p. 268.

60. Hans Moravec, *Mind Children: The Future of Robot and Human Intelligence*, Cambridge, Mass.: Harvard University Press, 1988.

61. Doel and Clarke, op. cit., p. 277.

62. Pierre Teilhard de Chardin was a Jesuit palaeontologist who predicted the unification of human consciousness and spirit in a single, enormous 'noosphere' or 'mind sphere'. The noosphere network would embrace the earth, marshalling planetary resources in a world unified by love. This echoes Hegel's philosophy of the birth of Spirit as the immanent purpose of historical change. (See Pierre Teilhard de Chardin, *Phenomenon of Man*, trans. Bernard Wall, New York: Harper, 1959.)

63. Heim, op. cit., p. 39.

64. E. McLuhan and Zingrone, op. cit., p. 259.

65. Heim, op. cit., p. 37.

66. See Peter L. Berger and Thomas Luckmann, *The Social Construction of Reality: A Treatise in the Sociology of Knowledge*, London, New York, Victoria, Toronto and Auckland: Penguin, 1991.

67. Slavoj Žižek, *Tarrying with the Negative: Kant, Hegel, and the Critique of Ideology*, Durham, N.C.: Duke University Press, 1994, p. 44.

68. Coyne, op. cit., p. 269.

69. McLuhan, op. cit., 1970, p. 116.

70. Marshall McLuhan, *Verbi-Voco-Visual Explorations*, New York, Frankfurt and Villefranche-sur-Mer: Something Else Press, Inc., 1967a, unpaginated.

71. Coyne, op. cit., p. 373.

72. E. McLuhan and Zingrone, op. cit., p. 296.

73. Levinson, op. cit., p. 18.

74. McLuhan, op. cit., 1987, p. 48.

75. Levinson, op. cit., pp. 37–8.

76. Ibid., p. 33.

77. Ibid., p. 112.

78. E. McLuhan and Zingrone, op. cit., p. 276.

79. McLuhan and Stearn, op. cit., p. 329.

80. Donna Haraway, 'A Cyborg Manifesto: Science, Technology and Socialist-Feminism in the 1980s', in *Simians, Cyborgs and Women*, London: Routledge, 1989.

81. Sherrie Turkle, *Life on the Screen: Identity in the Age of the Internet*, London: Weidenfeld and Nicolson, 1995, p. 268.

82. N. Katherine Hayles, 'Embodied Virtuality', in *Immersed in Technology: Art and Virtual Environments*, Mary Anne Moser (ed.) with Douglas MacLeod, Cambridge, Mass. and London: MIT Press, 1996, p. 3.

83. Heim, op. cit., p. 47.

84. E. McLuhan and Zingrone, op. cit., p. 249.

85. McLuhan, op. cit., 1987, p. 16.

86. Marshall McLuhan, *The Gutenberg Galaxy*, London: Routledge and Kegan Paul, 1962, p. 32

87. Genosko, op. cit., p. 11.

88. Marshall McLuhan and Harley Parker, *Through the Vanishing Point: Space in Poetry and Painting*, New York, Evanston and London: Harper Colophon Books, 1969, p. 31.

Select Bibliography

Paul Benedetti and Nancy DeHard (eds), *Forward Through the Rearview Mirror: Reflections on and by Marshall McLuhan*, Ontario: Prentice Hall Canada Inc., 1997.

Paul Levinson, *Digital McLuhan: A Guide to the Information Millennium*, London: Routledge, 1999.

Gary Genosko, *McLuhan and Baudrillard: The Masters of Implosion*, London and New York: Routledge, 1999.

Philip Marchand, *Marshall McLuhan: The Medium and the Messenger*, New York: Ticknor and Fields, 1989.

Marshall McLuhan, *Understanding Media: The Extensions of Man*, Cambridge, Mass. and London: MIT Press, 1994.

Eric McLuhan and Frank Zingrone (eds), *Essential McLuhan*, London: Routledge, 1997.

Websites
http://www.mcluhan.utoronto.ca
http://www.ctheory.com

Glossary

The medium is the message

The human's use of any communications medium has an impact that is of more relevance than the content of any medium, or what that medium may convey. The process of being in a virtual environment, for example, has a greater effect on our existence than the program in which we are immersed. The act of watching television has had a greater impact than what is shown on the television.

The rear-view mirror

When society and the individual are confronted with a new situation, they will attach themselves to objects of the recent past. We therefore perceive the present through a rear-view mirror. New media, including the car and the computer, are initially looked at in terms of previous technologies, such as the horse-drawn carriage and the typewriter.

Visual and acoustic media environments

When information is simultaneous from all directions at once, the culture is auditory and tribal. Although these cultures may superficially appear to be attached to the ear and mouth, for McLuhan the crucial criterion is that the media environment in which individuals communicate must have simultaneity and instantaneity of communication. Thus, even apparently visually-oriented media such as newspapers are in fact more aural in their presentation, because they provide a sense of information coming from everywhere in the world at once and being arranged in a non-linear fashion on the page. Virtual technology is thus an acoustic medium, for it has a high degree of tactility, immediacy and 'all-aroundness'. It is a medium of the 'ear' rather than the eye.

Visual cultures, on the other hand, substitute an eye for an ear. The eye has a point of view, which the ear does not. It internalises speech through writing, and separates individuals through print. It favours a linear, abstract format – commonly called text – and thus compels individuals to understand the world using a principle of 'one thing at a time'. The Western world has been dominated by the visual order, 'with procedures and spaces that are uniform, continuous and connected'.[1]

Hot and cool media

The basic principle that decides whether a medium is hot or cool is the degree to which that medium extends one sense in 'high definition'. In other words, it is the extent to which one of our senses is supplied with a lot of data. A photograph has more information than a cartoon, and is therefore a hotter medium. The telephone is a cool medium because the ear receives little information. The user therefore has more participation in a cool medium than a hot one.

The issue becomes more complicated with a 'meta-medium' such as the computer, in which a variety of hot and cool media (RealVideo and e-mail, for example) are superimposed and juxtaposed. In this context, virtual reality is arguably a very cool medium indeed.

Tetrad

This sets out McLuhan's four laws or effects of media: ampli-fication, obsolescence, retrieval, reversal. What aspects of society does the new medium enhance or amplify? What aspects of media that were dominant before the arrival of the medium in question does it eclipse or render obsolescent? What does the medium return to prominence from previous obsolescence? And what does the medium reverse or turn back into when it has run its course or been developed to its fullest potential?

Note

1. Marshall McLuhan and Harley Parker, *Through the Vanishing Point: Space in Poetry and Painting*, New York, Evanston and London: Harper Colophon Books, 1969, p. 1.

Acknowledgements

Thanks to Toby Clark, Duncan Heath, Cristina Mateo, Ron Delves, Fran Lloyd, the staff and students in the Faculty of Design, Neal White, and De Geuzen Institute in Amsterdam.